I0409486

Table of Contents

What is a worm farm?

Worm farming, otherwise known as vermiculture (vermis from the Latin for worm) is the process of harnessing earthworms to convert organic waste into the world's most nutrient-rich fertiliser; worm manure. Worm manure – also worm castings or vermicompost – is teeming with minerals, nutrients and beneficial micro-organisms essential for healthy plant growth, root development and disease suppression. Due to the nutritional superiority of worm manure, farmers and gardeners often refer to it as 'Black Gold', with one tablespoon enough to feed a small plant for three months. Before you get started worm farming, it is important to learn how it works. At its core, worm farming is a process designed to generate nutrient-dense compost. This compost is ideal for home gardens while being far easier to attain and less expensive than other composting methods. Traditionally, gardeners who want to use compost must either let it slowly decompose – which can be smelly (and time-consuming!) – or spend money on fertilizer made somewhere else, which can quickly add up (especially for those who like to do a lot of gardening). Worm farming is a great way to reduce your household food waste. As long as you have an area in your garden that is well sheltered like a balcony, verandah, pergola etc, it's easy to do and the benefits are well worth it. Some even like to have their worm farms indoors like a laundry or garage.

Here are some interesting worm facts:

Worldwide, approximately 6,000 species of earthworms are described in 20 families, eight of which are represented in Australia. Australian natives are estimated to total 1,000 species belonging to three of these families.Invertebrates make up 97% of species on earth without backbones and worms are just one of them. Worms are most definitely a gardener's friend and are vital to soil health. As they burrow beneath the ground, they consume soil, feeding on decomposing organic matter such as roots and leaves, sand grains and microorganisms. As the organic matter passes through their digestive systems, vital minerals and nutrients are transported and it's been shown that not only is worm digested soil healthier, it also has more phosphorous.Worms are a sign of healthy soil. If there's no food they'll go elsewhere. The more worms in your soil, the more nutritious it is, not only for them but for your plants! If you don't see any signs of worms, simply add more organic matter and they will find it.The burrows that worms create act as ducts that water and oxygen can pass through, helping to keep soil moist and aerated, vital for good plant growth allowing roots to grow and develop. Worms are hermaphroditic (they have both male and female reproductive organs) but they need to mate with other worms to produce offspring. After they've mated, worms form tiny, grain sized cocoons that are buried. They can produce up to two cocoons a week, each containing 1-7 hatchlings. After a two to four-week gestation period, the baby worms emerge.Lots of animals like birds and chooks love to eat worms but there's enough for them in a healthy garden to cater for them all – there could be as many as 1,000,000 worms living in an acre lot of land. Worms don't have lungs, they breathe through their skin. If there's too much rain, worms will rise to the surface to breathe as they may become starved of oxygen in water drenched soil. However, light paralyses worms so if they're out in it for more than an hour, they can't retreat back into the safety of darkness and will die. During darkness worms often feed on the soil surface. If you head outside with a torch you'll spot them all diving back under cover when they see the light, just like a pool full of synchronised swimmers.

Worms have no eyes, ears or teeth but can have up to five hearts. Contradictory to popular belief, if you chop a worm in

half it won't grow again and will die…
twice.

Why Start Raising Worms

You might be surprised to learn the numerous benefits that raising worms can offer. Whether your concerns are environmental, financial, or horticultural, worm farming can provide significant value to those who are willing to put in the effort. Although there are many benefits that come with worm farming, according to our readers, these are the four of the most significant:

1. Reduce Household Waste

Every day, we produce huge amounts of waste in our homes. Everything from banana peels to old newspapers would normally add up to a lot of waste thrown away. In fact, it is estimated that Americans generate about 67 million tons of organic waste every year, and less than a third of that waste ends up composted. Part of the reason for this is that many Americans think composting is too difficult or not worth their time and effort. Whether you are interested in reducing the environmental impact of your household, or you just like the idea of reusing what would otherwise be garbage in a useful way, vermicomposting is by far one of the best ways to do so. In addition, not just organic waste can be used. Even paper waste (such as newspapers and cardboard) play a role in the composting process.

2. **Limitless Supply of "Black Gold"**

Anyone who has been gardening for a while knows that fertilizer is one of the most significant expenses in home gardening. Nevertheless, the payoff (both in terms of enjoyment as well as the financial savings of growing your own fruits and vegetables) almost always outweighs the costs. But, what if you could create all the fertilizer you needed for free? Wouldn't that make gardening that much more enjoyable?

The compost generated from vermicomposting is called "black gold" for a reason! The incredibly nutrient-dense material can turn even the most barren soil (the type of soil you'll often find in your backyard) into great gardening soil. Over time, this "black gold" can save you hundreds if not thousands of dollars in fertilizer money! That saves time, money, and the environment!

3. **Valuable Teaching Opportunities**

Vermicomposting is a great activity to teach your child as well. It is simple enough that most children are able to pick it up quickly, but also requires a level of responsibility that proves invaluable down the road. Vermicomposting is most effective when it is maintained for a few minutes every day. It can be a great alternative (or addition) to teaching your child responsibility through the care of a pet. Not only that, but vermicomposting teaches your child about the value of conservation. Raising worms introduces them to the fun and rewarding world of gardening in a way they understand. Think about it: many kids play in the dirt anyway, now you can make it a worthwhile endeavor!

4. Great Conversation Starter!

Worm farms are popular in many parts of the world. However, it's still a relatively new concept across much of the United States. If gardening is something that you enjoy talking about with your friends, vermicomposting can be an interesting and fruitful addition to your conversation.

Benefits of Worm Castings

Worm castings generate the best compost in the world. Not only are they made through a completely organic process, but they also provide the perfect balance of nitrogen, phosphates, potash, and all the other plant nutrients that your garden needs to thrive. Of course, not all worms are the same, and the Red Wiggler worm is by far the best worm for vermicomposting. The worm casings themselves are the result of the digestive process the worms go through. The material itself naturally mixes into the soil of the worm farm, producing the highly valued fertilizer. If you are worried that you're somehow disrupting the natural process by harvesting the worm castings, don't be! Studies have shown that worms don't actually thrive best in their own castings, so you'll actually be doing them a favor every time you harvest new fertilizer for your garden!

Some other benefits of worm castings include:

- Completely non-toxic and organically produced fertilizer alternative
- Less odor than other fertilizers

Studies indicate that fertilizer created with worm casings can have as much as 5 times more nitrogen, 7 times the potash, and 1.5 times the calcium of regular soil.

Is Worm Farming Difficult?

In a word: NO! Worm farming is actually incredibly easy to learn. That's why we encourage you to teach your children how to do it; it's a great skill that is absolutely manageable by even relatively young children. In order to create a worm farm, all you have to do is follow a few simple steps:

Build or find a container – The first step is to find a container (typically wood or plastic) to hold the farm itself. While it is possible to build a container yourself, commercially-produced containers are more reliable over time, and are certain to not interfere with the natural processes of the worm farm.

Fill the container with appropriate bedding – Before adding the worms, you'll want to line the container with wet newspaper (or another appropriate bedding), then add some simple soil, either that you purchased or some that you already have in your backyard. The container itself should be moist enough to keep the soil loosely packed, but not so wet that the worms are at risk of drowning. It is also helpful to add egg shells to the soil (if you have them).

Populate the farm with Red Wigglers – This part is fairly simple: simply add the worms to your container. Don't worry too much about putting too many (or not enough) into the farm, as the worms will self-regulate their population fairly quickly.

Provide the compostable material – Compostable material can be everything from food waste to the remnants of lawn mowing. The simple rule here is: if it decomposes naturally, it can go in the farm. However, there are some specific types of compost that should not be put into your worm farm. These foods include: dairy, meats, citrus, spicy foods, fats, oils, and heavily-processed foods.

Maintain and harvest the farm – While it is important not to overfeed your worms, keep in mind that your worms like to eat a lot. In fact, Red Wigglers eat about half their weight every 24 hours, which means you can add new food for your worms every day (this is a great job for kids!). Remember to cut the food into the smallest pieces you can. Also, avoid putting dairy and meat into the farm. These are harder for the worms to digest and can create a significantly worse smell.

What Do I need to Start Worm Farming?

Vermicomposting is great because it doesn't require much of an investment, and provides significant dividends very quickly.

Worm composting bin

While you can technically use just about any container as a worm composting bin, some are far better than others. Commercially-made containers make it far easier to maintain new layers of your farm, and will be far more reliable than something you make yourself. We have a few worm bins available in our online shop that are optimized for the environmental concerns involved with worm farming.

Live red wiggler composting worms

As we mentioned, not all worms are ideal for vermicomposting. Red Wigglers are by far the best option because they generate the ideal blend of nutrients for gardening. While it is unlikely that you can find Red Wigglers in your backyard on their own, they are incredibly affordable and available on our website. For the cost of a bag of fertilizer, you can have the creatures you need to create your own fertilizer in perpetuity.

Worm composting accessories (optional)

As far as required components, you will be able to make do with just a bin, worms, and the raw materials your worms will eat. However, there are a number of accessories that make worm farming easier, and help increase your yield (effectively paying for themselves over time). Having the tools you need to ensure your farm has the right temperature, moisture, and pH prevents mistakes and saves you money down the road.

Choosing a Worm Farm

You can buy or build a worm farm, and they come in all shapes and sizes to suit all tastes and requirements. Most worm farms consist of a set of stacked trays with legs, and don't take up much room at all. They are ideal in size for a small household. If you're after a larger capacity worm farming system, one that can process large amounts of food waste, you can make one out of a recycled old bathtub or buy one of the commercial wheelie bin worm farms. These larger worm farms are ideal for places that generate lots of food scraps, such as larger families, schools, cafes, restaurants or workplaces in general. It's important to choose a worm farm that will fit in your available space that can cope with the volume of food waste you produce. It's important factor to consider the 'footprint' – how much space the worm farm takes up on the ground. As you can see in the picture above, my three worm farms occupy a fair bit of space, luckily it's in an unused corner of the backyard. I started with one worm farm , then added one more to cope with extra kitchen scraps, and then was given one – it wasn't planned to run three side-by side!

A bathtub worm farm has a capacity of around 200L, and as you'd expect, it occupies the same space as a bathtub, which is a fair bit of space!

A wheelie-bin worm farm can have a capacity of 140L, 240L or 360L and occupies very little space on the ground. It and has the advantage of being moveable because in has wheels. Consider your requirements when purchasing, and you'll be well rewarded, as worm farms last a very long time and are a great investment for the organic gardener!

How Does a Worm Farm Work?

Worm farms use earthworms to break down organic matter such as food scraps to produce worm castings and the liquid 'worm wee', properly termed worm casting leachate.All earthworms can do the same work, converting organic matter into valuable worm castings, but some breeds do a better job than others, so naturally, we choose the best worms for the job! The earthworms used in worm farms are in fact compost worms, which are different to the regular earthworms found in garden soil. Compost worm are surface feeders and don't burrow deep into the soil like garden earthworms do. The various breeds of compost worms, such as Tigers, Reds and Blues, are capable of eating their own body weight in food each day, so a kilogram of worms will consume that much food daily! By comparison garden earthworms only eat around half their body weight each day, so they aren't as good at composting lots of material really quickly, as it takes them twice as long. It's important to keep in mind that compost worms won't survive in your garden soil. Being surface feeders, they can't burrow deeply into the ground to the cooler soil in the heat of summer like regular earthworms, so they don't survive for long. They also need thick layers of composting organic material on top of the soil to feed on, so if there's no organic matter over your soil that is breaking down, they won't have any food. All earthworms are part of an ecological class of organisms called decomposers, they eat rotting organic material and turn it into worm castings. Since they don't have any teeth, earthworms need to wait till their food start to break down before they can begin to eat it. If their food is chopped up or broken up, it breaks down faster, and the worms can eat it sooner.

Now that we've covered basic earthworm theory we can now look at worm farm designs.

Worm farms are usually made from two stacked trays:

- The top tray contains the worms and food scraps.
- It has a lid to keep pests out, with air holes in the lid so the worms can breathe.

- It has drain holes in the bottom which allow any liquid to drip out.
- The bottom tray collects the liquid that drips out of the top tray.
- It has a tap or outlet on one side where the liquid can be collected.

The liner that sits at the bottom of the top tray stops the worms from falling through the drain holes into the liquid below and drowning. In commercial worm farms a piece of cardboard or some newspaper works fine as a liner. In homemade worm farms, where the drain holes are bigger, use a piece of shade-cloth or window screening first, then put the cardboard or newspaper down over it.

The worm bedding is where the worms live, it is usually just a damp layer of coconut coir, but you can use other materials such as damp shredded newspaper, or well-aged compost or manure.

The food scraps are kitchen scraps and other materials that the worms can eat – we'll go into detail of what you can and can't feed your worm later on in this book.

The cover which is also called a 'worm blanket' is an old hessian sack or a whole newspaper, it helps create a dark, cool and moist habitat by providing a cover over the bedding and their food, which encourages them to move into the food and eat it. The cover will eventually break down after a few weeks as it is biodegradable.

This is the basic worm farm design, but there are some variations. Many worm farm that you buy will have two or more top trays, the idea being that when the top tray fills up with worm castings, you can just add another tray to the top of it, and start putting kitchen scraps in the upper tray, and the worms will move up to the food. That also lets you remove the lower tray of completed castings to use in the garden while the worms are busy eating in the new tray. Lets have a look at another worm farm design and see how it also works.

How to Set Up a Worm Farm

When setting up a worm farm it is important to choose the right location. Worms like a cool environment, so if you locate the worm farm in a shady spot outdoors where it will not overheat from exposure to direct sun, your worms will be happy. You can place the worm farm on a shady side of a fence, at the side of the house where there isn't much sun, under a tree, or even inside a shed or garage as long as it doesn't get really hot in there during the summer. A protected spot on a balcony will work just fine too. It's also important to set up the worm farm in a location where you can easily get to it, so it's ideal to locate it close to your kitchen, which will be the source of kitchen scraps for your worm farm. If you can't easily access your worm farm you're less likely to use it. The Permaculture design principle of Relative Location explains how to set up a closed-loop sustainable gardening system with a worm farm and kitchen garden It's basic stuff, but the increased crop productivity, and long-term benefits of vermicompost are undeniable. Soil conditioned with this "black gold," is what keeps many farm and garden operations from going under. It improves soil structure, increases yield and even improves the taste of fruits and vegetables, and makes them last longer in the field.

TIP: Keep a small bucket or container with a lid in the kitchen to throw your food scraps into, and empty it into the worm farm when you're done. Lining with a piece of newspaper helps keep the container clean and the newspaper will be broken down in the worm farm with the food scraps.

Here are the basic steps to getting a worm farm stated:

- Assemble purchased worm farm or construct one yourself (see instructions for building a worm farm here).
- Place the liner (about 0.5cm (1/4") of newspaper or a sheet cardboard) in the top tray.
- Prepare bedding material (by either soaking coconut coir or shredded newspaper in a bucket of water till it is damp, or acquiring a container of well-aged compost or manure) and put into the top tray above the liner.
- Place the cover (worm blanket) made out of a whole damp

newspaper or a damp hessian sack over bedding.

- Add worms onto the bedding under the worm blanket cover, begin with around 500-1,000 worms..
- Allow a few days for the worms to adjust to their new environment.
- After a few days, begin to feed the worms lightly.

What You Need:

Worms: Eisenia fetida, are the most common type of worm used for vermicomposting. These worms are sold by the pound at many gardening centers or bait shops. You don't need a lot to start a home worm bin. One pound of these guys is equivalent to 1,000 worms. They reproduce like crazy and regulate their number based on the amount of food available.

Two plastic bins: We used 18 plastic storage bins with snap on lids. The box you use needs to be at least 12" deep. Make sure they're opaque. Worms like it dark.

A drill

A small flowerpot or a brick

Some old newspapers and household food waste

Putting it all together is easy. Master Vermicomposter shows you how:

1. Mark out holes on one of the bins. Using a pencil, mark out a series of holes around all four sides of the top of the bin. Mark out about 20 holes in the bottom of the bin. Leave the other bin blank. Take one of the lids and mark out enough holes so

that the bin will get some air exchange. We made our hole pattern for the lid in the shape of a worm.

2. Drill out the holes. For the lid and sides we used a 3/32" drill bit. For the bottom holes, we used a larger 3/16" bit.

3. Stack your bins. Put a brick or flowerpot in the undrilled bin and stack the drilled bin on top. This allows some space for the liquid to drain out of the top bin into the bottom one.

4. Prepare the bedding. Elliot says that the bedding materials are like "browns" in garden compost. Shredded newspapers work great, as does torn up corrugated cardboard. A few dried leaves work too. Just avoid anything with glossy color printing or leaves with a lot of volatile oil or strong scent. Once your bedding is in place, wet it down until it's the

consistency of a wet sponge. It should be moist, but fluffy.

5. Lay out some worm food. Table scraps are the best. Just don't add any oil or animal products like bone, meat or fat, or any dairy like butter or yoghurt. Citrus is okay, and Elliot says that the blue mold that naturally occurs on citrus peels is actually good and it inoculates your bin with beneficial substances that help your worms do their work. Just go moderately with acidic substances like citrus and coffee grounds. "Diversity is the key," she explains.

6. Add the wigglers. Once your bin is all set, bury a small amount of food scraps and let your worms loose on it. Worms naturally go for the dark, so they'll bury themselves in your table scraps. Don't worry, they usually can't find their way

out of the bin and escape. They don't want that anyway, and neither do you.

7. Tuck them in. To avoid fruit fly infestation, and worm escapees, take a few sheets of wet newspaper and lay them flat on top of your bedding. Then take a few more wet sheets and roll them up. Tuck them around the corners to form a seal so that everything stays in place and your worms are protected.

8. Put them to work. Don't expect much in the first few weeks. They are getting over the trauma of a new home. Once they're up to it, though, they can consume up to their own weight in food a day. So, if you put in roughly 1 pound of worms, try putting in just about a pound of scraps a day. Don't worry if you put in too much or too little, just make sure you add a variety of food scraps, so that the little guys will

have something to munch on. You can feed them every few days, or as infrequently as once every two weeks. Just make sure you replace the food that is disappearing. You'll see that some foods break down quickly (like ripe fruit) and others take forever (like root veggies and cabbage). To avoid bad smells, bury your food scraps underneath some bedding and vary the location of the food throughout the box.

9. Harvest your worm compost. Once the worms have done their work, you will see vermicompost in the bin. It's dark brown and looks like coffee grounds. To get some, without using fancy machinery, lure the worms to another area of the bin with fresh food. In a few days most of the worms will be working the new area, so you can carefully scoop out the finished compost. It's okay if you have a few

worms hanging on. Just make sure you leave most of them in the bin to keep working. You can also detach the top bin and pour out the "juice" that accumulates in the bottom bin. This stuff is like a high-energy drink for your plants. Dilute it or aerate it and feed your houseplants.

How to Build a Bath Tub Worm Farm

Large scale worms farms made from recycled bath tubs don't have multiple trays, just a single level. The bathtub is supported off the ground on bricks or it sits in a wooden stand or frame, and a bucket is placed below the drain hole to collect the liquid. The drain hole is covered with mesh or screen so only the liquid comes out.

The bottom of the bathtub is filled with coarse gravel for drainage, then the bath is lined with shade cloth above the gravel and then filled with bedding material. The rest of the layers are just like any other worm farm.

Since bathtubs don't come with lids, a timber sheet or wooden cover is used to protect the worms and keep them shaded.

What to Feed Your Worms

Firstly, what you feed your worms is important. There are some things that worms won't eat, and there are other things that are just simply unhygienic to put into a worm farm.

Remember that a worm farm is a vermicomposting system, it is used mainly for food scraps, which break down very quickly. Woody garden prunings and green waste are slow to break down and are better placed in a regular compost bin instead.

Let's have a look at what you can and can't put in your worm farm:

Things You Can Put In Your Worm Farm

Fruit & vegie scraps

- Bread & cheese
- Cooked vegetables, grains, pasta & rice – basically all vegetarian foods, no meat-based sauces!
- Coffee grounds & tea bags – as long as teabags are paper and not plastic mesh

- Egg shells – great source of calcium, a mineral which worms require in their diet to stay healthy
- Newspaper & unprinted cardboard (soaked) – no glossy printed pages

Things You Can Put In Your Worm Farm (With Caution!)

- Vacuum cleaner dust – only if your carpets are natural fibre, not synthetic carpets!
- Citrus & onions – only small amounts or none at all!
- Pet waste – only in a dedicated worm farm for pet waste only

Things You Can't Put In Your Worm Farm

- Fish & meat – this will stink and attract vermin such as rats, use a Bokashi bin instead to compost meat
- Garden waste – too slow to break down in a worm farm, put into compost bin instead
- Glossy and bleached paper – this is toxic, you don't want this anywhere near your garden
- Fresh manures – many animals are treated for worms with vermicides, which pass into the fresh manure and will kill your worms, compost them first!

How to Feed Your Worms

Secondly, how you feed your worms is also very important. Place the food on the bedding, beneath the cover, also known as the 'worm blanket', which is just a damp old hessian sack or a whole newspaper, fold it to fit if necessary. When you first set up your worm farm, add a small amount of food, and as the worms begin to feed in a few days, then add more. Don't overfeed your worms as the food will remain uneaten, and will begin to rot, which doesn't create a healthy habitat for your worms. When feeding the worms in the worm farm, don't cover the whole surface with food, place the food to one side only, and try to cover half of one side at the most. Just in case the worms don't like what you've just put in there, they can go to the other side of the worm farm where there's no food. If you cover it completely they'll have nowhere to escape to if they don't react well to your latest food offering. Once that half covered with food is eaten, add more food to the other side, and alternate sides, so there's always a food free side for them to move to if they need to. The worms in the worm farm will breed and the population of worms will grow. As they multiply they will eat food faster, and you'll be able to add more food. The number of worm will eventually self-regulate to match the size of the worm farm and the food available, so after a while you'll know how much food they can process.

How to Collect and Using Worm Farm Leachate

Worm casting leachate, aka 'worm wee', liquid gold for your garden!

Worm farms all have a tap or outlet to collect the liquid that seeps out from the worm farm, this liquid is often called 'worm wee' or 'worm pee' but the correct name is worm casting leachate.

To collect the leachate:

If your worm farm has a tap, just put a bucket under the tap, and turn the tap, the liquid will flow out to fill your bucket. When you've collected it all, close the tap, it's that simple.

If your worm farm doesn't have a tap, and just has an outlet (a pipe) for the liquid, then place a small bucket permanently under the outlet to collect the liquid as it's produced.

Even if your worm farm has a tap, you can permanently leave the tap open and place a bucket under the tap, this will prevent flooding! This is how I prefer to do it.

To use the leachate, always dilute it with water first before you use it on your garden, as it may be too strong to use directly. Always dilute it 10:1 with water, that means one part leachate to ten parts of water. When it is diluted it will be the colour of weak tea.

It's important to keep in mind that the leachate is it is not a fertilizer like worm castings, it's more of a soil conditioner that improves the health of the soil as it's full of minerals and beneficial microorganisms. Think of it more as a vitamin tonic for the plants and the soil, rather than a food. Since it's rich in beneficial beneficial microorganisms, you should always dilute it with rainwater, because tap water is chlorinated and will kill all the good bacteria in there!

How to Collect and Use Worm Castings

The worm castings, or vermicompost, is ready to collect when the bedding material and the food in it has broken down well and all that remains is a dark, rich, fine, moist substance, in which you can no longer see the food scraps.

A good time to collect the castings is in spring and autumn, because this is a good time to fertilize your garden.

The trickiest part of collecting worm castings is separating the worms from the castings! You want to keep your worms in the worm farm. There are a few techniques for harvesting worm castings which allow you to separate the worms out which we will look at in detail below.

How to Separate Worms from Castings

The 'Rainy Day' Technique

If you have a worm farm which uses stacked trays, you can wait for a day when it looks like it is about to rain. On rainy days the barometric pressure in the atmosphere drops, and the worms sense this, so they rise to the top to avoid drowning when the rain comes down. This is a natural survival instinct that for when the rain floods their burrows and tunnels in the ground. When they worms come to the top, they will all leave the lower tray, and will gather in the top tray or inside of the lid. When they all come up, you can quickly lift out the lower tray and put it aside for later use. Don't leave it out in the rain though, as it will become overly waterlogged, and lots of beneficial bacteria will get washed out, put it undercover somewhere and use it in the garden as soon as possible after the rain has passed. With worm farms that have a door at the bottom to harvest castings, simply harvest the castings and put them in a bucket for later use, but don't leave them sitting in the bucket for an extended period of time, as there's no drainage and any moisture at the bottom may become stagnant water.

The 'Pyramid' Technique

So what happens if it isn't going to rain anytime soon and you need castings? Well, there are other habits that worms have which we can take advantage of – worms dislike light, so make sure you don't expose them to direct sunlight when caring for them, they sunburn easily! To separate the castings from the worms, gather your castings, which will contain worms, put on some rubber gloves, and place a pile of castings on a low flat container or board, and shape it into a pyramid shape. Do this in a shady spot outdoors. The worms will not stay in the narrow pointed tip, and will burrow downwards to escape the light. Scoop off the tip of the worm casting pyramid, and put that into a bucket. Then reshape the pile into a pyramid with a new tip, and harvest the worm-free castings again. As you keep on doing this, the pyramid will get smaller and smaller, and the worms will keep moving to the bottom. When you have a small, low, flat pile full of worms, put it back in the worm farm.

The 'Let The Worms Decide' Technique

We can take advantage of yet another of the worms natural instincts to facilitate the harvesting of castings. When their bedding turns to castings, they will be basically living in their own waste, which is not their preferred environment. They have a preference for fresh bedding material with a supply of food. If you push the castings to one side of the worm farm to make a space to lay down fresh bedding material, put fresh bedding in in that space, and only lay food on the fresh bedding side, the worms will move over to the area with fresh bedding and food, and will move away from the side which contains only their waste (castings) and no food. Once all the food is finished in the castings, they will decide to move out to the nicer side, and you can then collect the castings! This technique works well in long, wide worm farms such as bathtub worm farms. In the process of harvesting worm castings, you'll find earthworm eggs or cocoons. They're easy to identify, they're small amber or yellow eggs about 3mm (1/8") in size which look like little beads but when you have a closer look at them they're shaped like tiny lemons. Pick these out when you come across them and return them back into the worm farm. Try to use the castings fairly soon after you've collected them, don't let them sit around for a very long time, and don't let them dry out, as they'll lose their beneficial value. Now that you've collected your castings, there are many ways that you can use them.

How to Use Worms Castings

In the garden – dig into the soil or place under mulch

Sowing seeds – add worm castings (up to 25% of total volume) to your seed raising medium

Indoor plants – add to the potting mix during growing season

Compost activator – add some worm castings to your compost bin to inoculate it with beneficial bacteria, which will help kick start your compost

Worm casting tea* – made similar to compost tea, full of nutrients and beneficial soil organisms, which can be used as a foliar spray on the leaves or watered into the soil.

Basically, you can use worm castings the same way you would use any slow-release organic fertilizer, it's that simple!

* Note – How to make worm casting tea (or compost tea) is a process that would take a short article to describe so I haven't included the instructions in this article!

Solving Common Worm Farm Problems

Worm farms are quite problem-free and easy to look after, but there are a few things to keep in mind that will make caring for your worm farm much easier.Here's a few of the biggest problems you might face and some simple solutions.

Protecting Worm Farms from Rain

Unless your worm farms is undercover, it will get rained upon, and some rainwater will get in, depending on the design. This is a benefit in my mind as it flushes out the castings and makes a good supply of worm casting leachate (worm wee) that you can use in your garden.

If your worm farm has a tap, and the tap is closed, then your worm farm may get flooded! The simple solution is to leave the tap permanently open and place a small bucket underneath. This will also prevent the tap getting blocked too.If you do need to leave the tap shut, you can save any worms from drowning if they fall into the liquid in the bottom by giving them and 'island' they can climb onto. Just place an upside-down terracotta pot into the bottom of the worm farm where the liquid collects. The terracotta pot is heavy enough so it won't float and move around, and the surface is not slippery like plastic, which will allow the worms to climb the sloped sides Also, remember to open the tap and collect the liquid from your worm farm once a week, and if your worm farm is exposed to the rain, collect the liquid immediately after it rains too.

Protecting Worm Farms from Hot Weather

The fastest way to lose all the worms in a worm farm at once is to accidentally let them get cooked in hot weather! We've already discussed the location of the worm farm earlier in this book, it should be in a protected, shaded location away from direct sun. Sometimes, even the shadiest location might get direct sun exposure during summer, because the sun is directly overhead, or because the hot west afternoon sun comes in from the side as the sun lowers in the evening. Worm farms can overheat simply due to the high temperature of a hot summer's day, because the air outside is hot, and they're in an enclosed container. To alleviate this problem, prop the lid the worm farm open a bit to let air circulate through and to release any hot air that may be building up under the lid. You can simply lie a stick across the top of the worm farm and place the lid over it so there's a gap between the top edge of the worm farm and the lid. On really hot days, you may need to cool down the worm farm by watering it with a watering can and rainwater. Make sure there's some form of cover material (the 'worm blanket') in place such as newspaper over the bedding and food, to keep the worm's environment dark and moist. Open the tap to prevent flooding and place a bucket underneath to collect the liquid. Water with a watering can making sure you evenly dampen the whole surface. DO NOT use tap water if you can , it is contains chlorine, which will kill a lot of the beneficial bacteria on your worm farm! After wetting down the bedding and cover material, the water will slowly evaporate to create a form of evaporative cooling which will help the worms cool down.

The best way to protect worm farms from direct hot sun is to cover them with a screen of some sort that is light coloured and will reflect the sun, with sufficient air-space underneath it so it doesn't trap the hot air over the worm farms and cause them to overheat. You can remove the screen in the cooler seasons, and put in in place during the warmer seasons.

I've found that cheap reflective plastic tarpaulin sheets, suspended high above the worm farms to allow air to flow underneath, work extremely well. You can tie the bottom of the tarp to a brick or other heavy object to stop it flapping around in the wind.

Dealing With Insects in the Worm Farm

It's natural to have a few other insects in your worm farm, but some are unwelcome guests!

Ants do not belong on your worm farm, and it may be because it's too dry in your worm farm, as ant's don't like moist environments. To discourage ants, dampen down the worm farm with a watering can full of rainwater, and to stop them getting in there, create a barrier, an 'ant moat' by sitting the legs of your worm farm (if it has legs) in tray of water. The water will evaporate on hot days so remember to keep it topped up. I've seen suggestions of using taller narrow containers filled with oil which won't evaporate but in my mind that will create a disgusting mess as dirt gets blown in by the wind. I reckon water is a much tidier solution!

Smearing a band of Vaseline around the legs of a worm farm is another suggested solution but it's likely to melt in hot climates. Ants aren't a problem unless you overload your worm farm with lots of sugary food, and if the food has been there long enough to attract ant's there's a chance you may putting in much more food than the worms can eat. Remember to chop up the food so breaks down faster!

Vinegar flies are those tiny flies that fly up into your face when you lift the lid on a compost bin or worm farm. They are attracted to rotting food, especially fruit, as are fruit flies, so the best way to prevent them breeding is to cover the food scraps beneath a damp newspaper (remember that all important 'worm blanket' cover over your bedding!)

Other insects such as millipedes are not a problem, they are decomposers and feed on rotting organic matter, returning the nutrients to the soil. Slaters, also known as wood lice, pillbugs, roly-polys or butcher boys are also beneficial decomposers and are in fact land-based crustaceans! Soldier fly larvae, which look like giant silver-grey maggots are also beneficial, though they look a bit creepy. Springtails are unmistakeable little insects which hop around on the surface when you lift the lid, they are also beneficial, they're all part of the decomposer community too, a natural part of the Earth ecosystems recycling processes.

If you see tiny white worms in your worm farm, they are not baby earthworms, these worms are entrachyadids, they are not harmful but do indicate that your worm farm has become a bit too acidic. Correct the acidity by sprinkling a small amount of garden lime, dolomite or wood ash (which are all alkaline) in your worm farm every few weeks.

Odour, and How to Fix a Smelly Worm Farm

A healthy worm farm will have little to no smell, perhaps a faint but pleasant earthy smell just like healthy soil or a forest floor. If it has a sharp vinegar smell it may be too acidic, add crushed eggshells, garden lime, dolomite or wood ash to correct the problem. If it smells quite offensive, it is an indicator that the system has become quite anaerobic from too much uneaten food. To fix this problem, stop adding any more food, add a sprinkling of garden lime, dolomite or wood ash, and lightly stir up the existing food scraps to aerate them on a regular basis. Once the smell disappears, then begin feeding the worms again.

Ready to start your worm farm?

Great! The easiest approach is to purchase a pre-made farm online, but it's a more expensive alternative to creating one of your own.

When building a worm farm of your own, choose a container that's suitable for your available space. Here's what you'll need, according to the U.S. Environmental Protection Agency:

- Two rubber or plastic bins — one taller bin with a lid, and one shorter, bottom bin without a lid
- A drill to drill holes in the taller bin
- Scrap of screen (The EPA says window screens work fine provided there is no metal)
- Waterproof glue to keep screens in place
- Shredded, non-coated paper, moistened in water to create the right environment
- Worms, of course!
- Trowel or tool to move compost around in the bin
- Food scraps container to collect your household scraps (They can be fed directly by putting food scraps

right in your worm farm, but feeding them once a week is preferable)

From here, drill small holes about two inches from the top of the taller bin on both sides, along with four 1/8-inch holes near the bottom corners.

Next, you'll want to cover each hole with vinyl screening, glue it in place, and let dry completely.

Then, place the tall bin inside the short bin (which should not have any holes). Now you're ready to add your paper, soil, and just enough water to dampen everything, adding the entire mixture into the tall bin. Fill the bin about three inches deep with this mixture. Now it's time to add those worms! The EPA recommends letting them get used to their new environment for a day before feeding them your food scraps.

Create a feeding schedule for your new worm farm.

As mentioned, while it's okay to toss your food scraps directly into your worm farm as you create them, it's recommended that you hold onto them and feed your farm just once a week. Before you feed your worms, throw in a healthy handful of the shredded paper. Then add your food scraps on top of the paper, and cover the paper and food scraps with dirt and additional, pre-moistened paper. There's a logic here! "Exposed food attracts fruit flies, but covered food scraps don't. Add dirt and moist paper to the bin until the worms have made enough compost to use to cover the food scraps." From there, close up your bin, and let the worms work their magic.

And remember — there are 'yes' and 'no' foods for your worms.

Your worms won't be particularly picky, but we tracked down some general guidelines to keep in mind.When feeding your farm family, a good rule of thumb is to say "yes" to things like non-citrus fruits, vegetables, egg shells, bread scraps, coffee grounds, used tea bags, and wait for it even vacuum cleaner dust! It's best to avoid things like citrus fruits, onions, garlic, meat, fish, poultry, eggs, dairy, and processed food.